MEDITATIONS ON
Butterflies

A COLORING & HAND-LETTERING
ACTIVITY JOURNAL

Kristen D'Angelo
Artwork by Maryjo Koch

GIRL FRIDAY BOOKS

Butterflies are a breath of beauty fluttering by. They are mystery chronicled upon wing; they bring forth the grace and wonder of the world to our eyes every day. — KRISTEN D'ANGELO

For centuries, men, women, and children, the young and old alike, have found themselves admiring the graceful flight of the butterfly. This aerial display, however, is only the beginning of this winged beauty's allure. Once the butterfly draws you in with its visual charms, more wonders of its world begin to be revealed, forever captivating your spirit.

The butterfly's domain is one of mystery and grace, where the miracle of metamorphosis rules. The tiniest egg becomes a colorful caterpillar. That caterpillar then changes into a curious chrysalis, from which emerges a butterfly. As an admirer of these soaring sensations, you will discover their magical world unfolding into yours.

Butterfly Families

Have you ever really watched a butterfly dance with the wind or waltz upon a flower top? Often the beauty of a fleeting butterfly may be noticed for just a moment in the midst of an otherwise busy and full life. Let's take some time to be with the butterflies.

Flying flowers come in all sizes and every color of the rainbow. Being with butterflies begins with learning the six butterfly families: the Swallowtails, Brush-Foots, Whites and Sulphurs, Gossamer-Wings, Metalmarks, and Skippers.

Swallowtails *(Papilionidae)*

In the hours of the ambered afternoon, one might spy members of the magnificent Swallowtail family *Papilionidae* gliding by. These breathtaking beauties are a sight to behold as they move through fields of gold, flaunting their splendid tails of hue.

Swallowtails are large, striking butterflies whose family, *Papilionidae*, includes over 40 species in North America and 600 worldwide. They get their name from the tail-like extensions on their hindwings. They are a favorite for many butterfly lovers as their size and brillant features make them easy to spot and enjoy.

Brush-Footed Butterflies *(Nymphalidae)*

Sometimes travelers or mysterious migrators, Brush-footed fliers from the family *Nymphalidae* dot the earth's skies with color. These lovely wanderers are beloved by many as they illuminate the heavens with their vivid hues. They are not just butterflies but also brilliant wildflowers in flight.

Nymphalidae is the largest family of butterflies with more than 5,000 species worldwide and about 230 in North America. Many species are medium- to large-sized and radiantly hued; however, the underwings are, in contrast, often dull and can look remarkably like dead leaves or bark. These butterflies are called brush-footed because they are known to stand on only four legs, while the other two are curled up.

Whites and Sulphurs *(Pieridae)*

Some of Mother Nature's tiniest creations, White and Sulphur butterflies from the family *Pieridae* may be petite in size, but when they dance in skies of painted blue, their beauty is a thing to behold. Sit in the garden awhile and enjoy a bit of their harnessed sunshine.

The *Pieridae* family has about 1,000 species worldwide and approximately 70 in North America. Most pierid butterflies are white, yellow, or orange, often with black spots. The word "butterfly" is believed to have originated from the Brimstone, which was called the "butter-coloured fly" by early British naturalists.

Gossamer-Winged Butterflies *(Lycaenidae)*

Like tiny jewels floating upon the breeze, the Gossamer-Winged butterflies of the family *Lycaenidae* move brightly through the skies of North America. Small yet exquisite, to catch of glimpse of these stunning treasures is truly a luxury!

Gossamer-Wings are small butterflies, usually under a 2-inch wingspan. They are dazzlingly marked and sometimes flaunt a metallic gloss. In North America, there are about 140 species and around 7,000 worldwide. Catch the wind and let your heart soar with these delicate-winged butterflies.

Metalmarks *(Riodinidae)*

Over summer blossoms, the midnight hour approaches as the Metalmark glides by, for these butterflies' stunning iridescence is nothing short of a starry night upon wing.

A mini kaleidoscope of wings flutters by as the Metalmarks of the family *Riodinidae* dot the meadows. The hues range in these butterflies from a transparent shimmer to muted red-orange grays to iridescent blue-greens. Small copper and silvery metallic spots also shine to give them their common name of "metalmarks." There are approximately 1,300 species of these butterflies worldwide and about 30 in North America.

Skippers *(Hesperiidae)*

Dressed in *rosa* and *azul*, Skipper butterflies adorn the garden as rainbow mists of morning dew.

Named for their quick, darting flight habits, Skippers are often overlooked but when spied are entertaining and fun to watch. These zippy fliers have small to medium wingspans averaging from about 1 to 1½ inches. More than 3,500 species are recognized worldwide and about 300 are found in North America. Skippers also have generally stocky bodies and large compound eyes. Their wings are usually small in proportion to their bodies.

By gardening for the benefit of butterflies, you will not only bring some of their beauty and wonder into your yard, but also help to ensure the survival of some of these lovely creatures! Below are a few simple guidelines to follow when creating your butterfly garden.

Creating a Beautiful Butterfly Garden Habitat

Host Plants

A butterfly garden must have host plants. Butterfly host plants provide a place for butterflies to lay their eggs and in turn serve as the food for the caterpillar (larval) stage of the butterfly. Most species of butterflies will only lay their eggs on one or a select few species of plants. Without ample food for the caterpillar, chances are the species will diminish in quantity.

Different regions have different species of butterflies. In order to maintain and hopefully increase the population of butterflies, you should find out which butterflies are found in your area and help those species by planting the correct host plants. Check with your neighborhood library or nursery to find a list of your local butterflies and their host plants. Planting a butterfly garden filled with plants native to your area and other butterfly-attracting flora will not only benefit the butterflies that visit your garden but also will provide food for birds, insects, and other wildlife.

It is important to note that most host plants are native plants, which typically grow in the wild. It is unlikely that you will see many of them around your city's typical landscaping. In many areas when land is groomed for development, it often destroys the natural habitat where other native plants grow. No native host plants, no butterflies.

By planting butterfly host plants in your garden, you are helping butterflies to thrive and also ensuring that they will continue to grace our world for generations to come.

Plant with Your Environment in Mind

When deciding what to plant in your butterfly garden, take some time to learn more about the benefits of native plants. No matter where you live, by choosing to include some native plants in your landscaping plans, you'll make things easier on the environment. Because your local native plants are already ideally suited for your region, they need less upkeep and maintenance. They usually require less water, so by choosing native plants you are helping to conserve that precious resource as well.

Native plants provide a beautiful, hardy, drought-resistant, low-maintenance landscape that benefits the environment and all the lovely creatures that reside there!

Nectar Plants

Once you have drawn butterflies into your garden with a variety of desirable host plants, you can keep them coming back by providing an ample supply of their favorite nectar-producing plants. Nectar plants contain sugar-rich liquid produced by their flowers, which in turn attracts butterflies, nectar-loving birds, and pollinating insects. Butterflies have a natural relationship with flowering plants. Just like bees, butterflies pollinate flowering plants, and in return for their services, they are provided with an energizing nectar drink.

As flowering plants and butterflies have evolved together over the course of millions of years, they have learned to talk to one another by speaking the "language of flowers." Some plants will continually change the colors of their blooms to show the butterflies which flowers still have nectar in them. For this reason, butterflies can be partial to certain hues such as yellow, red, purple, or pink. These colors are frequently advertised on flower heads, which are rich with nectar.

Other plants have evolved to provide easily accessible landing pads for butterflies. Butterflies prefer these flat-topped flowers because they are simple to set down and feed upon. They also provide a clear view to any possible predators hiding within.

Check our suggested list of butterflies' favorite nectar plants (www.obsessionwithbutterflies.com). Notice how many of them flaunt blooms that are flat on top and are attractively colored to appeal to butterflies.

Butterflies are God's confetti thrown upon the Earth in Celebration of Love."
—Kristen D'Angelo—

Family Lycaenidae ~ Gossamer-winged Butterflies

No Insecticides or Pesticides

It is important that you should take care never to use chemical insecticides or pesticides in your butterfly garden. Instead, practice organic gardening, which is safe for caterpillars and other beneficial insects.

Food and Shelter

If you really want to create the ultimate butterfly paradise, you can place or hang several shallow dishes in your yard. Fill one dish with overripe fruit and the other with damp mineral-rich mud and stones.

Male butterflies will especially love these nutrient-filled treats, and females, too, may drop in for a sampling if they're not too busy laying eggs. Butterflies favor overripe fruit such as plums, apricots, peaches, grapes, watermelons, and cantaloupes.

Wooden butterfly houses also add a nice touch to any butterfly garden, and they provide shelter for butterflies at night or during cooler weather. Beautiful hand-painted butterfly houses can enhance a garden with their charming designs.

Create It and They Will Come!

Believe it or not, butterflies have great memories. Once they have identified your yard as a butterfly haven, a place where their favorite nectar and host plants can be found, they will become frequent and regular visitors.

You will be able to sit back and enjoy nature at its best. A show—130 million years in the making—will play out in your yard almost every day. With an all-star cast of eggs, caterpillars, chrysalises, and butterflies, it will be sure to be entertaining and fascinating for all who come to watch.

The Butterfly's Life Cycle

Did you know that there is no such thing as a baby butterfly? Butterflies do not start out as babies; they hatch from eggs as caterpillars. From there, they take an incredible journey to become an adult butterfly.

Egg

A mother butterfly takes her eggs very seriously. In fact, "egg laying" is really the main objective of her short, sweet life. She is so serious about this because over the course of millions of years, evolution has taught her that if she does not choose the right spot for her egg, it probably won't survive.

To predators, such as wasps, flies, beetles, and ants, the egg is a tempting little packet of nutrients just waiting to be eaten. In addition to various predators lurking about, the eggs may also fall victim to harsh weather and a vast array of viruses, bacteria, or fungi.

A typical egg develops in three to ten days, an eternity for an immobile embryo the size of a pinhead. About the only natural defenses butterfly eggs have are color and camouflage. A green egg on a green leaf is invisible; a brown egg becomes a speck of dirt; a bright red or yellow egg signals predators: "Go away! I don't taste good!"

Most female butterflies will only lay their eggs on one or two specific species of plants. Even when a prospective plant feels, tastes, and smells just like the right species, it may be rejected by the female for a variety of reasons. It may be too wet or too hot. Evidence of competition may also be present. Chewed leaves mean less food available for her young when they hatch.

On average, a female butterfly visits up to ten host plants before picking the perfect one. Depending on her species, she may lay a single egg or a cluster of eggs. She may lay them on the underside of a leaf, in a crevice, or on the tip of a twisted tendril. Location is everything to her—an investment in her species' future!

Caterpillar

When an egg begins to darken in color, a caterpillar is ready to emerge. Just the size of a dash or a comma, this tiny crawler enters the world by eating its way out of its shell. With all major body parts intact from the onset, this miniature muncher is ready and able to do what it does best. Eat!

It is thought caterpillars consume more vegetation than all other insects combined. In fact, a typical caterpillar gains over three thousand times its body weight. In human terms, this would be like a ten-pound baby becoming a three-thousand-pound adult.

No doubt Mother Nature designed caterpillars to be efficient eaters; she did not, however, design them to be efficient warriors. A newly emerged caterpillar faces all the same dangers it did as an egg and more.

Newly interested predators including birds, lizards, and spiders are among the large list of dangers a caterpillar faces. Such luxuries as claws, beaks, and venom are not at a caterpillar's disposal. Nor can a caterpillar scamper, hop or fly away quickly from a perilous situation.

Warriors? No, caterpillars are not warriors! I would like to think of some of them as munching magicians, sly masters of metamorphosis and illusion. Some caterpillars choose to look like a vein in a leaf, while others go for the bird-dropping look. Better yet, why not fool your audience entirely by making them think you are a green snake with large daunting eyes, and then pop out orange glands, which emit a foul odor? Whatever your method, hopefully your show will be a success, and the threatening predator will depart.

Some other tricks caterpillars pull out of their hats are to feed at night or on the underside of a leaf. Clip off the leaf they have munched and tidy up your grass—maybe no one will notice they are around.

Chrysalis

As a caterpillar prepares for its final act, to "pupate," or change into a chrysalis, it will often wander away from its host plant. It may search for a safe haven, a solitary spot, away from other caterpillars and any predators they may attract. A very small percentage of species pupates under soil or leaf litter in a crude cocoon. Most butterfly caterpillars, however, become rock climbers. They spin a silken line and "girdle," or harness, themselves to a stick or other chosen spot.

Once securely in place, they perform their magic once again. This is the finale, the highlight of the caterpillar's show. The total transformation of caterpillar to chrysalis is amazing.

Whichever form a chrysalis takes, it is designed to camouflage and confuse predators. This is an intentional case of false identity. The butterfly forming within will hopefully be mistaken by predators for something else.

On average, a butterfly will pupate in its chrysalis for four to fourteen days. In extreme cases some species have been known to hide out for six to seven years! Hopefully, its true identity will not be revealed to passersby.

If a predator were to look closely at a chrysalis, it might see the outlines of wings, antennae, or an abdomen. They might even see that it is an immobile butterfly in the making. It's a neatly packaged food source, nutrient rich and easily accessible. If a chrysalis's secret identity is revealed or found out and a predator attacks, the situation is usually a grim one. Defenses are few.

Through the use of illusion, camouflage, and manipulation, a few fortunate chrysalises survive longer than most of their kind. Within their seemingly lifeless structures, miracles are occurring. Soon these patient pupas will be released to experience a whole new world.

Adult Butterfly

When a chrysalis cracks open, a newly formed butterfly emerges. Its wings are wet and crumpled, but soon they will begin to dry and unfurl. The butterfly will be free to fly off into a different dimension. This new world contains many of the same dangers that the old one did, but the butterfly is now armed with a new and improved body.

On their wings, butterflies flaunt a lovely mosaic of scales. These scales absorb heat and also help a butterfly escape sticky situations such as spider webs. But most importantly, they help the butterfly to converse, or "wing talk." Communicating with friends and foe alike is something a butterfly must do often. Mother Nature has created some species that exhibit "false heads" complete with eye spots. Cleverly, these false heads are located on the hindwings, far from the butterfly's real head.

Dating Game

When not defending themselves or seeking nutrients to live on, butterflies spend most of their time participating in the "dating game." When a male is ready to court a female, he must find one. Some males do this by setting up a territory, which they guard to keep other males out. Others go out and actively search for their mates. Often-times they can be seen "hill-topping," or basking in the sun as they show off their colors, hoping to attract a female.

When a male successfully finds a female, he must then court her. In an aerial dance, he will try to sprinkle some "love dust," or pheromones, onto her antennae to make her fall in love with him. When the butterflies finally mate, they can be together for up to nine hours. Isn't love grand? Afterwards, the female and her fertilized eggs will be on their way. The male butterfly will go back to collecting nutrients for his next affair. This propagation of species is the business of butterflies—to love and be loved!

Acmon Blue

Love rose like a butterfly from my heart.

Whisper of Grace

FAMILY

Lycaenidae

PRACTICE

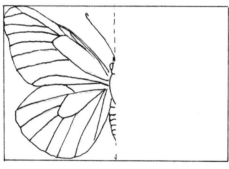

How To Draw a Butterfly

1. DRAW A BOX THE SIZE OF A BUTTERFLY - DRAW A LINE DOWN THE MIDDLE.
2. DRAW JUST ONE SIDE.

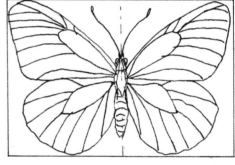

3. THEN WITH TRACING PAPER, TRACE OVER WHAT YOU DREW.   4. TURN THE TRACING PAPER OVER- LINE IT UP TO MATCH THE OTHER SIDE.   5. BURNISH THE IMAGE WITH A BONE FOLDER - YOU WILL THEN HAVE AN EXACT REPLICA OF THE OTHER SIDE.

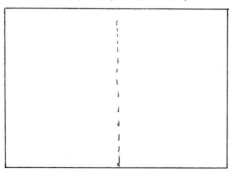

Practice Drawing Your Butterfly

USE A #2 PENCIL SUCH AS A BIC PENCIL 0.5mm.

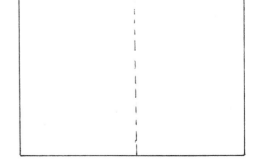

My Peace dwells on wings of butterflies.

In skies of Grace, blue butterflies dance, tiny wings of joy waltz as Heaven glances.

𝒜 *a*

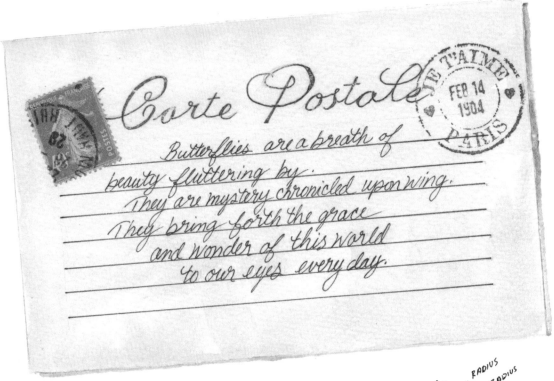

Carte Postale

JE T'AIME
FEB 14
1904
PARIS

Butterflies are a breath of beauty fluttering by.
They are mystery chronicled upon wing.
They bring forth the grace and wonder of this world
to our eyes every day.

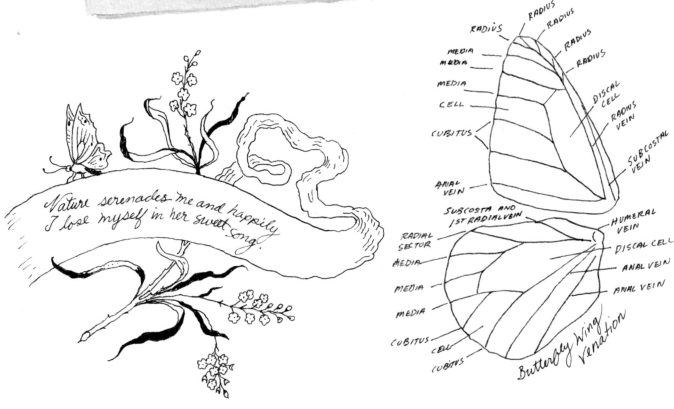

Nature serenades me and happily I lose myself in her sweet song.

Butterfly Wing Venation

RADIUS
RADIUS
RADIUS
RADIUS
RADIUS
RADIUS
MEDIA
MEDIA
MEDIA
CELL
CUBITUS
DISCAL CELL
RADIUS VEIN
SUBCOSTAL VEIN
ANAL VEIN
SUBCOSTA AND 1ST RADIAL VEIN
RADIAL SECTOR
MEDIA
MEDIA
MEDIA
CUBITUS
CELL
CUBITUS
HUMERAL VEIN
DISCAL CELL
ANAL VEIN
ANAL VEIN

Old stars of the deep, dark night, oh butterflies of the golden field,

teach me to be a whisper of grace, a soft echo tenderness yields.

_____

_____

_____

_____

_____

_____

_____

_____

_____

_____

_____

_____

_____

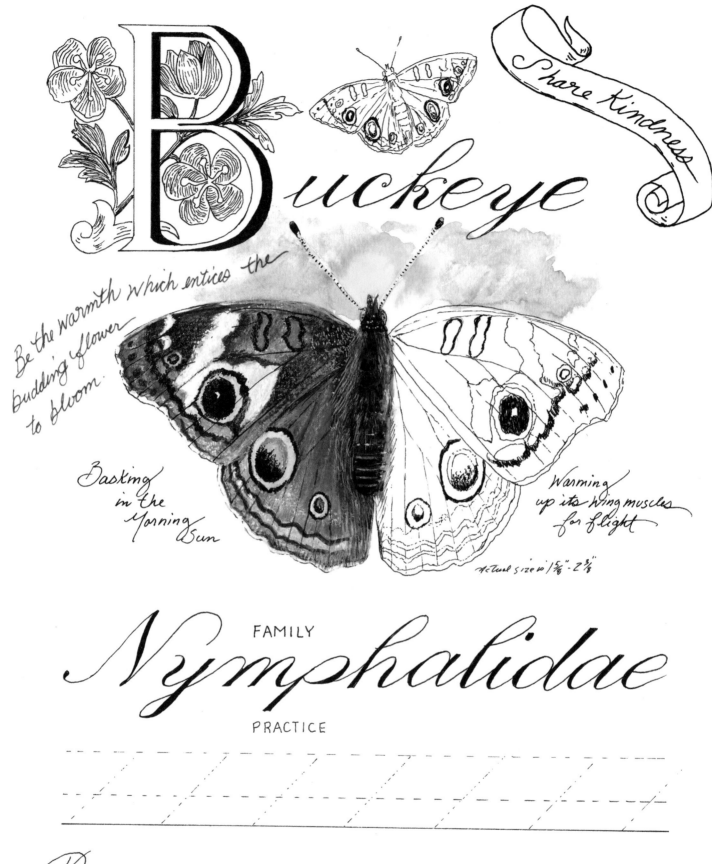

Share Kindness

Buckeye

Be the warmth which entices the budding flower to bloom.

Basking in the Morning Sun

Warming up its wing muscles for flight

Actual size w 1⅝" - 2⅜"

FAMILY

Nymphalidae

PRACTICE

Bask in the blessings of your day; let gratitude and kindness carry you away.

On a breath of bliss, be a wish, float as
dandelions dream and kiss.

$\mathcal{BB}$

Favorite nectar
plants for
butterflies:

Today I saw the beauty of:

FOREWING

THORAX
HEAD
ANTENNA

LABIAL PALPS

PROBOSCIS

PROLEG, MIDLEG, HINDLEG

HINDWING
ABDOMEN

Charity soared
with my
spirit in a
butterfly waltz.

AIR-MAIL

RÉPUBLIQUE FRANÇAISE
50c

My Gratitude List This Day:

Share Kindness

Kindness... May the world feel your kiss today! You change everything when you spread your wings of goodwill. Smiles are shared, care is spoken, joy nurtured, charitable moments created, Love chosen.

California Dogface

Also known as Flying Pansy

FAMILY
Pieridae

Play as the Light

PRACTICE

No garden truly blooms until butterflies have danced upon it.

Dreaming Bud, awaken to Love.

Cc

Ways to brighten another's day:

_____

_____

_____

_____

_____

_____

HEAD

THORACIC SEGMENTS

ABDOMINAL SEGMENTS

TRUE LEGS

SPIRACLE

PROLEGS

ANAL PROLEGS

CATERPILLAR ANATOMY

Carte Postale

I dreamt I was a butterfly...

I set a forest of trees ablaze in leaves of

amber bright... And moved wings

of butterflies to stained-glass flight.

In Nature today
I saw the Beauty of...

1. _____

2. _____

3. _____

4. _____

5. _____

6. _____

Play as the Light

Upon golden wing take flight, dance the skies as an untethered kite.

_____

_____

_____

_____

_____

_____

_____

_____

_____

_____

Diana Fritillary

Dream Come True

One of the loveliest Fritillaries

The Greek name for 'butterfly' is 'Psyche.'

The same word, 'Psyche,' means the soul.

FAMILY

Nymphalidae

PRACTICE

Divine creation can be seen painted on the canvas of a butterfly's wing.

All over the world butterflies soar into the hearts of man; they sail boundless and free, speaking the universal language of beauty.

$\mathscr{D}$ d

PRACTICE THE LETTER 'D'

Reflections

God dreamt a butterfly and
the wind whispered,
"Let love soar."

Dreams and Wishes:

POST CARD

How can I make my day brighten like a flower:

Dream come true

You are Creation's dream... so come true.

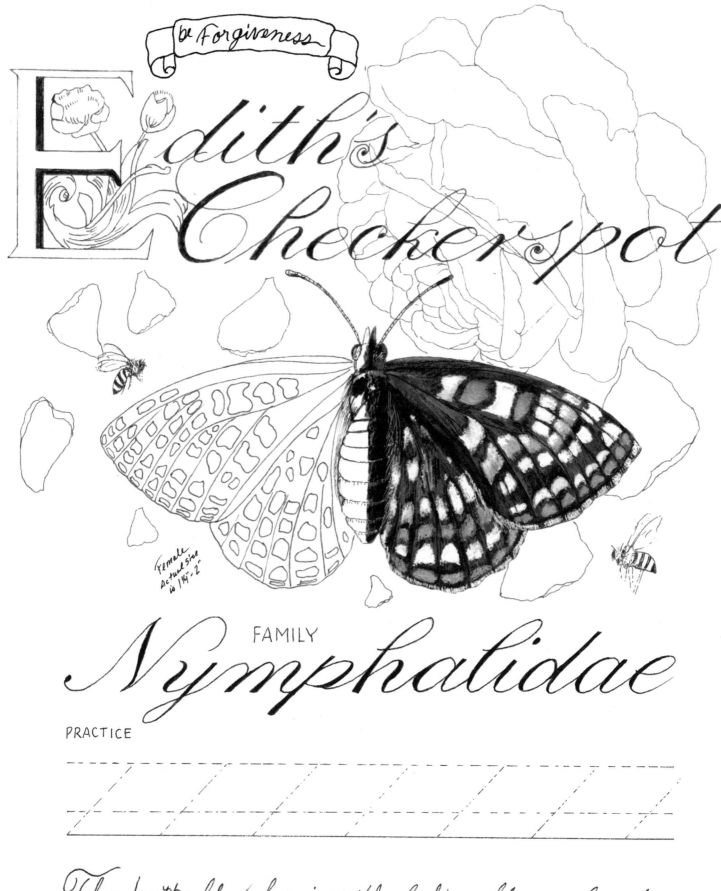

E is for Forgiveness

Edith's Checkerspot

Female Actual size is 1¼"-2"

FAMILY
Nymphalidae

PRACTICE

The butterfly forgives the fading flower, for she knows in its wilting it does promise another bloom.

Falling leaves, falling stars
Seasons come and go
Lifted by grace, falling in faith
Our wings forever know
Falling leaves, falling stars
Upon Grace, Love is ours

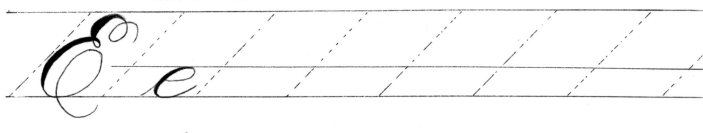

My Three Acts of Kindness Today

No. 1

No. 2

No. 3

Watch a butterfly in your garden, then write down your observations:

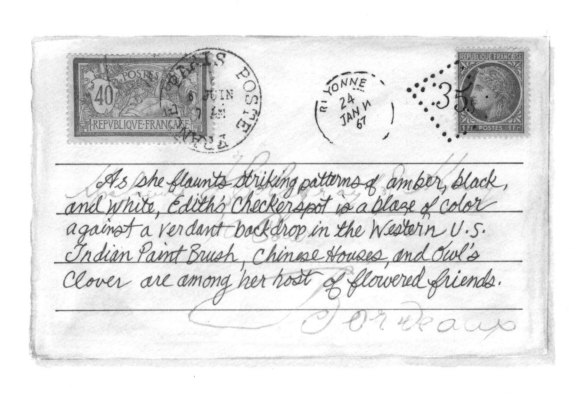

As she flaunts striking patterns of amber, black, and white, Edith's Checkerspot is a blaze of color against a verdant backdrop in the Western U.S. Indian Paint Brush, Chinese Houses, and Owl's Clover are among her host of flowered friends.

be Forgiveness

Cherish life as if it were a fleeting butterfly.

Life as a Soaring Blossom

Fulvous Checkerspot

water color sketch of host plant

Castilleja

A colony of Fulvous Checkerspots is always associated with Castilleja — their host plant.

Actual size is 1¼ - 1½

FAMILY

Nymphalidae

PRACTICE BY TRACING OVER THE LETTERS

Like the petals of a rose, slowly I unfold my life as a soaring blossom.

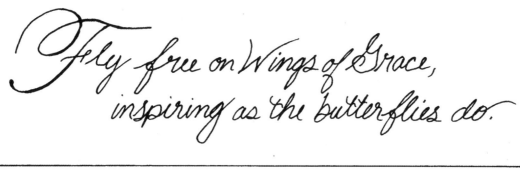

Fly free on Wings of Grace,
inspiring as the butterflies do.

_____

_____

_____

_____

_____

_____

_____

_____

_____

_____

_____

_____

_____

_____

DANCE OF THE BUTTERFLIES

Tiny miracles flutter by

Enchantment adrift upon the sky

Charming patterns of hue painted on wing

Melodies of seduction they do sing

Flowers, friend, and foe are engaged

Intrigue and fascination are waged

Spectators follow with captive eyes

The alluring dance of the butterflies

Watercolors
I used
for leaves
and stems:
OXIDE OF
CHROMIUM
by WINSOR & NEWTON
GREEN GOLD
by DANIEL SMITH

for berries:
CADMIUM RED
by WINSOR & NEWTON

for white
butterfly:
YELLOW OCHRE
by WINSOR & NEWTON
IVORY BLACK
by WINSOR & NEWTON
LAVENDER
by DANIEL SMITH

for light
orange
butterfly:
CADMIUM
YELLOW DEEP
by WINSOR & NEWTON
YELLOW OCHRE
by WINSOR & NEWTON
IVORY BLACK
AND INDIGO
by WINSOR & NEWTON

PRACTICE TRACING OVER THE LETTER 'F'

Life as a Soaring Blossom

Share Love's song, sing it up to the skies,
watch Angels dance as Heaven sighs.

_____

_____

_____

_____

_____

_____

_____

_____

_____

_____

_____

_____

Gray Hairstreak

Take a Stained-Glass Flight

Actual size is 1-1¼"

FAMILY
Lycaenidae

PRACTICE

Days of gray melt to spring on the warmth you bring.

The sun does not ask the moon —
Nor the flowers the butterfly.
Trees play with wind.
Clouds decorate the sky.

G g

Be kind to butterflies – give them a garden!

Butterfly Flower List

Butterflies are attracted to flowers by their nectar, scent, and color. Their favorite colors: purple and yellow.

Mod. CA = 79

PACCHI
PARTE
CENT.
30
SUL BOLLETTINO

20 para

Genova P.P.
DEPT. OF ENTOMOLOGY

Old Man Winter with his tufts of gray,
for both the flying and the still,
to Spring's bright flowers,
he gives way...

Butterflies do not like the color scarlet!

In spring She warms a Soul bright,
Moving Wings in stained-glass flight.

_____

_____

_____

_____

_____

_____

_____

_____

_____

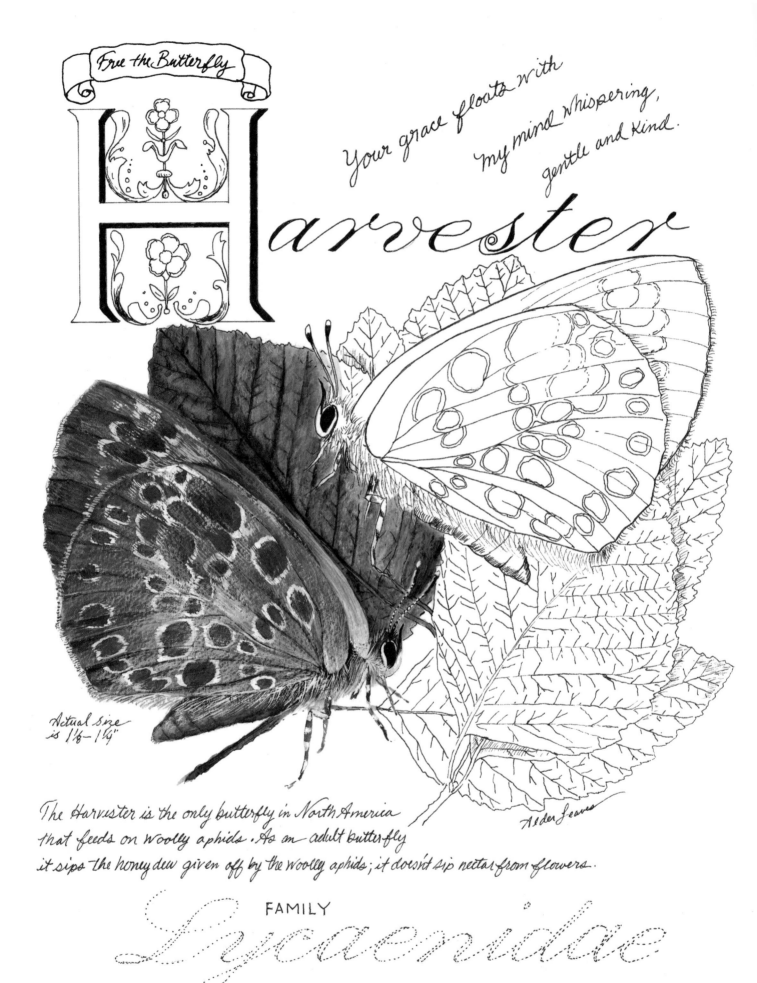

Free the Butterfly

Harvester

Your grace floats with
my mind whispering,
gentle and kind.

Actual size
is 1⅛ – 1¼"

Alder Leaves

The Harvester is the only butterfly in North America
that feeds on woolly aphids. As an adult butterfly
it sips the honey dew given off by the woolly aphids; it doesn't sip nectar from flowers.

FAMILY

Lycaenidae

PRACTICE BY TRACING OVER THE LETTERS

Soar like the spirited Butterfly,
who is as passionate upon tattered wing
as she was on her maiden flight.

H h

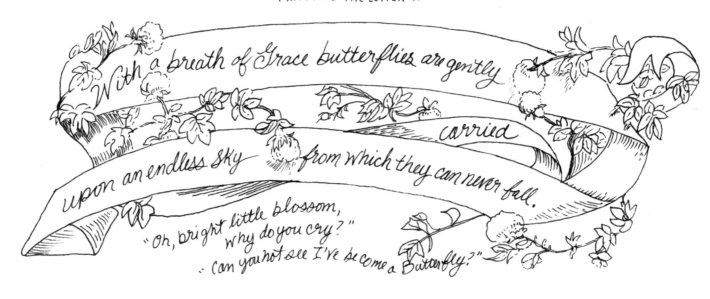

With a breath of Grace butterflies are gently carried upon an endless sky from which they can never fall.

"Oh, bright little blossom, why do you cry?"
"Can you not see I've become a Butterfly?"

Write how your butterfly garden touched your soul and transformed it.

The garden of my life...

Softly, I feel the flutter of your wing—in the garden of my life you sing.

Carte Postale

FINISH THIS THOUGHT: The spirit of the butterfly teaches us that no matter how many times our lives are destroyed, we can rebuild and renew...

Free the Butterfly

Release love, Let it take wing,
Unfold your heart, Free the butterfly.

Indra
Swallowtail

Actual size
is 2½ - 2⅞

FAMILY Papilionidae

PRACTICE BY TRACING OVER THE LETTERS

As a Butterfly, I sip the sweetness of Life, like nectar
for the spirit, inspiring my soul to joy and peace.

In a garden of hearts live,
Forever fluttering give.

I i

If you were a butterfly,
what colors would you be?

_____

_____

What patterns would be

on your wings? _____

_____

What would you write
in a letter to:
Dear World,

_____
_____
_____
_____
_____
_____
_____

POST CARD

Butterfly and Bee are not
about "me" or who is "better"
as they waltz a wild
flower together.

Flying Flower,
soar into the hearts
of others. What wisdom
you share....

Dear world... We treasure Peace on Earth!
Watch us seek, dream, awaken, discover, bloom, heal.
Love, Love, Love, the time is now to Change, Embrace & Soar!

Julia

This long-winged Brush-foot butterfly is a very rapid and agile flyer. It visits flowers for nectar. The caterpillars feed on passion flowers.

Actual size is 3/4 - 3 5/8"

Passion shines from all

FAMILY

Nymphalidae

PRACTICE BY TRACING OVER THE LETTERS

Watch with delight as a butterfly's beauty Reflects in the sparkle of your eye.

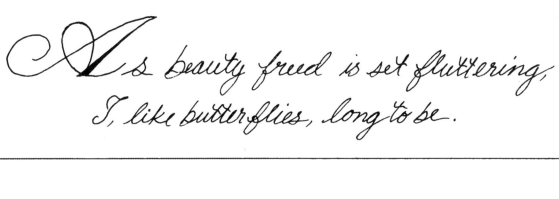

As beauty freed is set fluttering,
I, like butterflies, long to be.

J J j

List the
butterfly species
that live in your
vicinity:

_____

_____

_____

Tender moments shared
are the sweet
and burgeoning
buds of
spring.

What flowers do
you talk to in your
garden:

_____

_____

_____

List your
caterpillar
food plants:

_____

_____

_____

List your
butterfly
nectar plants:

_____

_____

_____

POST CARD

PINE
8
PA

REPUBLIQUE CENTR

CENOPHODES HYLAS

PROTECTION DES VEG

GOSSAMER-WINGED

BUTTERFLIES
Lycaenidae

SWALLOW-TAILS &
PARNASSIANS
Papilionidae

TRUE SKIPPERS
Hesperiidae

WHITES & SULPHURS
Pieridae

BRUSH-FOOTED
BUTTERFLIES
Nymphalidae

Butterfly
EGGS

Sweet life is borne on breezes of Spring. A tiny new bud, an egg left by wing.

Passion Shines from all

I wish to know the Spring of creation, Where
Passion shines from all I see.

King's Hairstreak

Petal Rose & Took Wing

A butterfly feeds on sugar-rich or proteinaceous fluids to renew energy reserves.

Actual size: 1 3/16 - 1 7/16"

Butterflies have feet which taste, and a built-in straw, or proboscis, which can be used to drink up sweet nectar.

FAMILY

Lycaenidae

PRACTICE BY TRACING OVER THE LETTERS

Softly, I feel the flutter of your wing in the garden of my life; you sing, and your grace floats with my mind, whispering gentle and kind.

As a glimmer of His thought
Soft petal rose & took wing
Borne of creative Passion
My spirit became Spring

K K

PRACTICE THE LETTER `K`

Flowers are the Keepers of
the Sun's warmth and Storm's
tears, and in their blooms
they share the Grace
with all who pass and
come to see.

Write about
where you find
comfort:

POST CARD

NEW YORK

10c
POSTES

Your comfort becomes the breeze
Which lifts and carries fallen leaves.
With the flowering of my heart I see
Your beauty soar silently.
So I fly the skies of time
With your wings next to mine.

Petal Rose & Took Wing

Let Nature & her gentle butterflies move Hearts who have forgotten how to soar.

_____

_____

_____

_____

_____

_____

_____

_____

_____

_____

_____

_____

Large Marble

Actual size is 1½-1¾

Finish painting the
Large Marble

The Marble Butterfly's wings are so exquisitely soft in pattern and color that they look as though they have been fashioned out of a fine stone. Look for this small flying masterpiece among the summer mustards of the Western United States.

Whites & Sulphurs

Family Pieridae

PRACTICE BY TRACING OVER THE LETTERS

Fragile wings carry bright dreams . . .
Wind never fails a vision that sails upon love.

_____

_____

_____

_____

_____

_____

_____

_____

_____

_____

_____

PRACTICE THE LETTER `L`

Fearful butterflies become bound,
as soar-less kites strung to ground.

Not far away life's miracles dance
With brave wings who faith chance.
Upon the breeze I dream to glide
Agape-winged, in Love abide.

The veins in the wings of
butterflies help keep the wing in the
correct flight position. The veins are
arranged to help identify which family of
butterfly a species belongs to.

Brave Wings Faith Chance

Dear Courage,

Help us stay true to you in Love. Let your strength shine on, life as in faithful flight, trusting wisdom, we hold each other in deep care— Our hearts will bravely answer your noble call.

Monarch

The Monarch's long migration is one of the most amazing feats in nature. They have been documented to fly almost 1,900 miles in a single season between Canada and Mexico.

Actual size: 3½ - 4"

FAMILY
Nymphalidae

Brush-foot

Join the Kaleidoscope of Wings

PRACTICE WRITING 'BRUSH-FOOT' AND THE LETTER 'M'

Mm

Butterflies are God's Confetti, thrown upon the Earth in celebration of Love.

Magical Monarchs

Kings and queens of the skies, Magical Monarchs soar over the lands as winged gems caped in orange beauty.

True treasures of nature, these bright royals need no crown nor sword to rule the hearts and move the spirits of their subjects.

For to catch a Monarch butterfly in flight is to be held by mystery, to waltz with Creation for just a moment, to be enchanted by Her fiery Love.

You are what you behold... as part of Glorious Creation ... you too can bring inspiration and joy to others...

Spread the warmth of the fiery monarchs ...
                    touch the world with kindness today.

POST CARD

Graceful magicians ... butterflies are all about transformation ... Name some ways you too can change or transform yourself to better inspire others. What journey might you take to migrate yourself closer to becoming Love?

Join the Kaleidoscope of Wings, where faith brightly sings.

N

Northern Crescent

Actual size is 1⅜ - 1½"

In the northeastern United States, among summer fields, where the golden aster grows, the Northern Crescent butterfly brightens up the forests with its vivid orange hues. It is not a butterfly but an amber wildflower in flight.

Family *Nymphalidae*

PRACTICE BY TRACING OVER THE LETTERS

N n

PRACTICE THE LETTER 'N'

Wander into the Garden, drawn by fragrant flower, song of bird and brook... Sit with a butterfly, breathe in Grace, and upon departing Beauty, long to return to Her again and again.

_____

_____

_____

_____

_____

_____

_____

_____

_____

_____

_____

_____

_____

Zinnia
Lavender
Rosemary

Bee balm
Heliotrope

Lippia

European Columbine

Sea Side Daisy

Marigold

15.

Butterfly seeks not solitude
but instead friendship in
flowers of gratitude
Or
Drawn by fragrant
flower into the garden
I wander...
Song of
bird and
brook and
Stream
How I long to return to
This .... dream

Images from: Mira Calligraphiae monumenta by Georg Bocskay

Aster
Coreopsis
Lantana

Violets

Double European Columbine

Wild Radish

Golden rod

Snapdragon

Gloriosa Daisy
Pincushion
Cosmos
Statice
Field Mustard

Summer Phlox

proboscis
Butterfly's Drinking
Straw to sip up
nectar
From flowers

Purple Coneflower

Red Valerian

Thistle

Lupine

Mallow

Monkey Flower

Friendship and Flowers

You are to me as a butterfly to a flower.

Old World Swallowtail

Actual Size 2 1/4" - 4"

As if flown from off a vintage canvas, the Old World Swallowtail paints the forest floor in classic beauty. Timelessly she sails the flowered waves of North America.

O o

PRACTICE THE LETTER 'O'

FAMILY

Papilionidae

PRACTICE BY TRACING OVER THE LETTERS

There is peace in knowing Grace in all Creation. Be cradled by the Universe, a space of indescribable beauty; know the infinite treasure of love, which both surrounds and is within yourself.

_____

_____

_____

_____

_____

_____

_____

_____

_____

_____

_____

_____

Take
the day. Remember
the glory that surrounds you.
Behold colorful Angels softly
fluttering... Watch with delight while a
butterfly's beauty Reflects in the sparkle of your eye.
Know only grace as she circles from above,
Gently embracing your heart.
Soar peacefully with her on   the wings of the
perfect love   that you are.

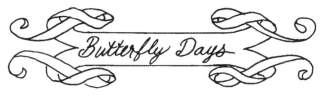

Petals of a rose choose unfolding...
I, like blossoms, long to love.

_____

_____

_____

_____

_____

_____

_____

_____

_____

_____

_____

_____

Painted Lady

On their wings butterflies flaunt a lovely mosaic of scales. These scales are most important for many reasons. They absorb heat, which helps to enable the butterflies to fly. They may also help a butterfly escape a sticky situation, such as a spider's web, as they fly off quickly and make for an easy escape.

Brush-Footed Butterfly

Most importantly, the butterfly's wings help the butterfly to converse, or 'wing talk'. Communicating with friends and foe alike is something a butterfly must do often.

Actual size: 2⅛"

Nymphalidae

PRACTICE BY TRACING OVER THE LETTERS

Catching rays, my colored wings fly high above in the blue sky.

Looking below at shadow's cast, I realize it's in the past.

Joy flashes from the gray, peace carries me away.

The butterfly flits from bloom to bloom, as not to miss one fragrant plume. In her sight exists no garden finale, only paradise of flowered hill and valley.

𝒫 𝒫 𝒫 𝒫 𝒫 𝒫

Wing!

Talk

After emerging from its chrysalis, the new world contains many of the same dangers that the old one did, perhaps even more, but the butterfly is now armed with a new and improved body.

It has:

Although contacting and flirting with a mate is of the utmost importance to a butterfly, staying alive takes priority. With the use of their 'talking wings,' butterflies send out clever messages, which are most effective in keeping predators at bay.

After Maria Sibylla Merian
engravings from 'Erucarum Ortus'

Use your imagination to color this in.

Proboscis
(PRO-BOSS-KISS)

• Six legs and a set of wings, which can fly.
• Compound eyes, which can see in every color and in every direction.
• Antennae, which can smell extremely efficiently.
• Feet, which taste, and a built-in straw or proboscis, which can be used to drink up sweet nectar.

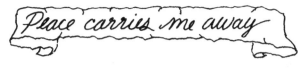

Peace carries me away

Butterflies grace the autumn fields, as sweet memories dance the meadows of our minds.

This Common Queen is found from Nevada and Southern California to Kansas and Texas, and then along the Gulf States to Florida and Southern Georgia.

Queen

Flies in successive broods from mid-spring to late fall, using milkweeds and related plants in the family Asclepiadaceae as caterpillar hosts.

Danaus gilippus
Common Queen
Actual size
is 2¼-3"

Catch a Flying Flower

Dance in dabs and strokes of every hue, move gracefully upon the canvas God created for you.

PRACTICE THE LETTER 'Q'

Q q

Tenderly planted seeds sprout love,
like an emerging dream we watch you grow.
As a bright flower in the garden of our hearts, forever
does your beauty stay, even as the bud flies away.

_____

_____

_____

_____

_____

_____

_____

_____

_____

_____

_____

_____

What's a Mother To Do?

On average a female butterfly may visit up to 10 prospective host plants before picking the perfect one. Even after she has made her choice, she may spend up to 9 hours surveying and selecting precise leaves on which to deposit her eggs.

A concerned mother plans for her children's future! She lays her eggs diligently. Depending on the species, she may lay them singularly or in a group. She may lay them on the underside of a leaf or in a crevice.

Most always she lays them on the most tender leaves. She may use the 'simple eye' on her abdomen to lay on the tip of a twisted tendril.

A minute spec of life is set upon a leaf, out crawls a miracle, a butterfly to be.

TO: The Nymphalidae Family

1 Brush-Foot Lane

Southwestern and Southeastern United States

Catch a flying flower, soar with love.

✱ Build a butterfly garden!

Catch a Flying Flower

Hope given wings, the butterfly has faith beyond fear, forever soaring with love.

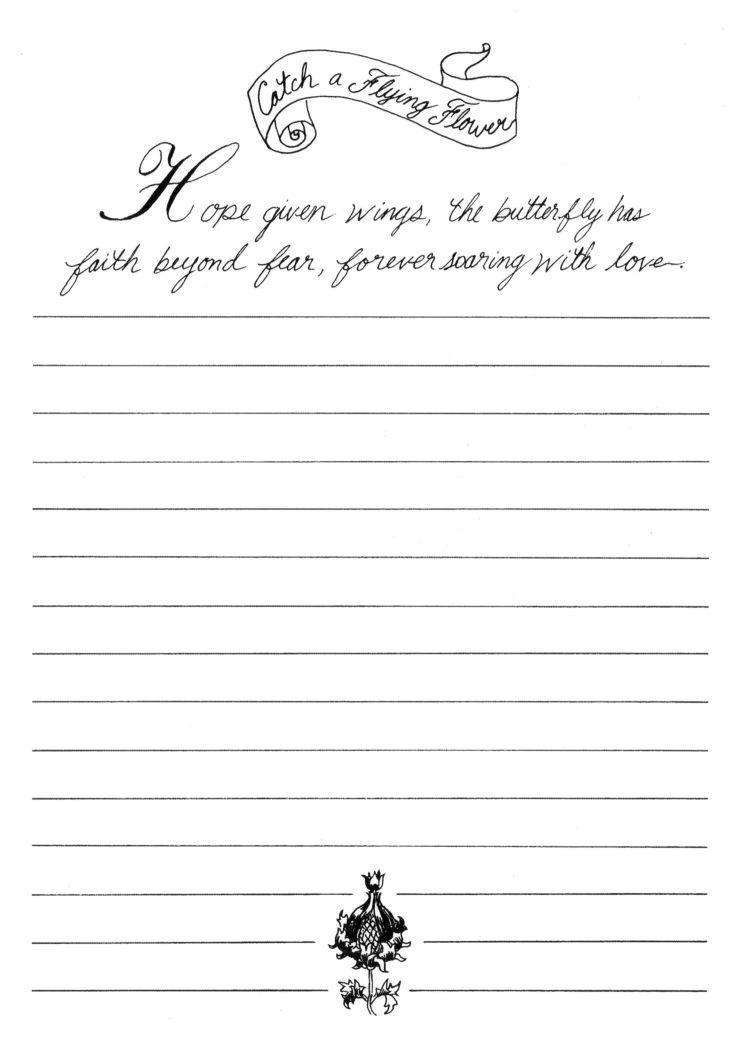

Red Admiral

Actual size: 2–3"

With a blaze of glory, Vanessa atalanta scouts the deserts and mountains of North America. Predators are cautioned by color, as spectators salute the fiery flight of the Red Admiral butterfly.

FAMILY
Nymphalidae

Brush-foot

PRACTICE BY TRACING OVER THE LETTERS

Shine Your Inner Light

Go out into the world as the shining
gift of Love you are.
Blaze bright, beautiful Little Winged Star.

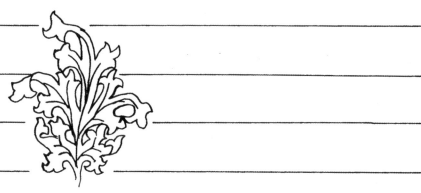

Shine Your Inner Light

You can see all the colors of a rainbow in the scales of the wings of butterflies.

R r

If butterflies had no scales, the wings would be very thin and clear as glass.

The scales on a butterfly's wing make the wing strong — they catch air, which helps the butterfly to fly easily and gracefully.

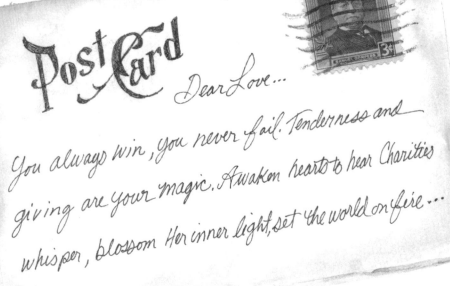

Post Card

Dear Love...

You always win, you never fail. Tenderness and giving are your magic. Awaken hearts to hear Charities whisper, blossom Her inner light, set the world on fire...

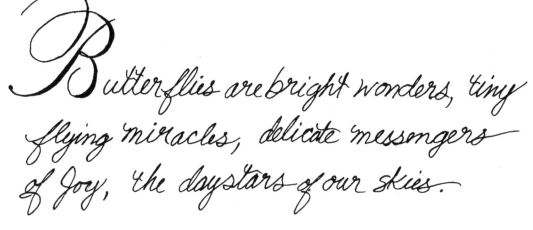

Butterflies are bright wonders, tiny flying miracles, delicate messengers of Joy, the daystars of our skies.

Sara Orangetip

Wake a Flower

When the sweet Sara Orangetip flutters by, you know that spring time has arrived in Southern California's coastal communities.

FAMILY

Pieridae

PRACTICE BY TRACING OVER THE LETTERS

Actual size: 1 3/8 - 1 7/8"

Echoes of charity move the world.
Peace is not self but others, pearled.

Be like Sun at noontime hour, share some warmth and wake a flower.

I heard a whisper of Grace,
An echo from above
Saying, 'Come little flower,
Awaken with Love.'

Some Ways

to touch the world with Kindness:

Be part of your community,
go for a walk with a neighbor,
help tend a community garden,
share a meal with someone
in need, be a friend
to a lonely heart ...

S s

CAROLINA POSTALE
CARTE POSTALE
POST KARTE

Nature is never solitary...
Awaken to the Universe's simple gift of the
butterfly. Watch with fascination and joy as a
jeweled treasure glides by and gently touches
your soul.
In the oak-dotted foothills of the Western U.S.
one might spy the Sara Orangetip gliding by. This
dainty beauty gently moves through the fields of
golden mustard, as a white petal upon a breeze.

PRACTICE THE LETTER 'S'

Wake a Flower

Love is kept by being given away, in Spring's innocence long to stay.

Tiger Swallowtail

Eastern Cottonwood

Eastern
Tiger
Swallowtail
Papilio glaucus

Caterpillars
feed on
Cottonwoods

FAMILY
Papilionidae

PRACTICE

A self-made sanctuary, a place for change and growth,

a chrysalis is made in anticipation of both.

In solitude, the butterfly steps from the chrysalis, the soul from the self. Butterfly and time cannot pass a garden together.

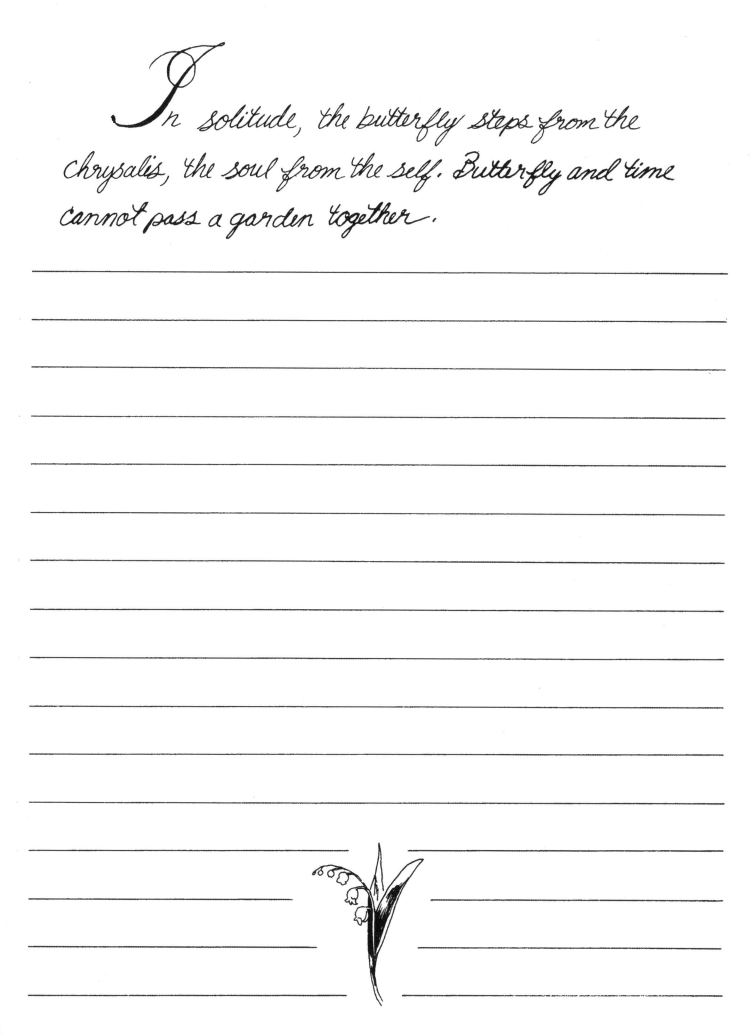

𝒯𝓉 𝒯𝓉

CARTE POSTALE

NEW YORK
MAY 19
4 PM
10036
NEW YORK

30 CENTIMES

As a caterpillar prepares its
final act, to change into a chrysalis, it may
search for a safe haven, a solitary spot. They spin a
silken line and harness themselves to their chosen spot.
Once securely in place, they perform their magic
once again. This is the finale, the highlight of the caterpillar's
show. The audience will be amazed as they watch the total
transformation of caterpillar to chrysalis, chrysalis to
Graceful Butterfly.

Forest cascades mirrored stream ~ Painting sails on butterfly wing
In dim stillness I change ~ cradled in silent slumber
Sun dawns as a golden dove ~
The treasure found in my heart
is love

Write your favorite quote:

_____

_____

_____

Change
in
Stillness

Treasure Is Love

The winds sang for Her, the sun danced,
color and form performed a magnificent duet.
If you listen carefully you might hear Creation whisper
to your heart and stir a movement of Grace.

Invitation to Dream

Umber Skipper

Actual size
1 ⅛ - 1 ⅜"

Skippers are named for their rapid skipping flight.

The Umber Skipper is a tiny traveler, a California native who dots the grassy fields in orangish black. This simple, sweet butterfly is often overlooked as he skips upon the yards and parks in which he flies.

FAMILY
Hesperiidae

PRACTICE BY TRACING OVER THE LETTERS

Be it Fairy or butterfly wing, both invite you to revisit fanciful days of yore, when you soared on your imagination to enchanted places, sparkling with the magic of Love.

_____

_____

_____

_____

_____

_____

_____

_____

_____

_____

_____

_____

Butterflies dance
on the
whisper of romance,
Alight upon
their wings,
Sun-kissed and joyfully
soaring into
Love's sweet
bliss.

PRACTICE THE LETTER 'U'

U u

Invitation to Dream

Close your eyes tight
Sprinkle a little Fairy Dust
Now open them to a Childlike Reverie
Where each twilight twinkle
Is an invitation to dream

Miracles awakened Imagination playing

Viceroy

The Viceroy mimics three butterflies: the Monarch, the Queen, and the Soldier.

ACTUAL SIZE: 2½-3¾"

We set forests of trees ablaze with the sun's loving gaze on Angels' breaths of Grace

The Viceroy caterpillars feed on the leaves of willows, poplars, cottonwoods, and their catkins.

FAMILY *Nymphalidae*

Brush-foot

PRACTICE BY TRACING OVER THE LETTERS

Vv

PRACTICE THE LETTER 'V'

Gentle rays of gold ignite
Pray keep these petals aflight
Guided by a compass of Hope

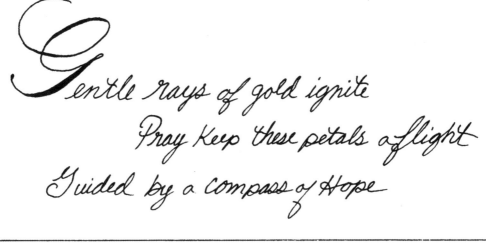

I am miracle awakened
Imagination playing
A single flying flower
Drawn by divine hours.

Anatomy of a Flower

PETAL
PISTIL
STYLE
STAMEN
FILAMENT
SEPAL
OVARY
OVULE
STEM

Is there someone who you would like to imitate? Why?

_____

_____

_____

Post Card

YELLOW SPRINGS
OHIO
DEC 24
1230P
1915

U.S. POSTAGE
TWO CENTS

Imitation is the sincerest form of flattery, but for the Viceroy and Monarch, whose patterns of orange-plus-black wing coloration are remarkably similar, it is a form of survival. Because Viceroys look like Monarchs, predators learn to avoid them as they do the Monarchs.

Miracles awakened, Imagination Playing

Joy of fragrant plumes Humbled from bloom to bloom — I dance with starlight upon wing.

Western Pygmy Blue

Lost in a Butterfly's wing

The smallest butterfly in North America, the Western Pygmy Blue, may be delicate in stature (only ⅜ of an inch), but when she dances, shades of blue flash and rival that of the oceans and skies, over and up in which she flies.

FAMILY *Lycaenidae*

PRACTICE BY TRACING OVER THE LETTERS

One could spend a lifetime lost in a butterfly's wing.

W w

PRACTICE THE LETTER 'W'

Meet me where the butterfly graces the gardens.

Tremendous beauty can be found in the tiniest of things... for who has ever thought to rival that of a Butterfly's Wing?

This tiny little Western Pygmy butterfly continuously flies throughout the year in Southern California and Southern Texas — migrating from these Southern U.S. colonies northward across the Great Plains and deserts of the west to eastern Oregon and Colorado.

CARTE POSTA

Sonoran Blue
Silvery Blue
Western Pygmy Blue
Echo Blue
Cassius Blue
Acmon Blue
Arctic Blue
Rea Kirt's Blue
Western Tailed Blue
Eastern Tailed Blue

UNITED STATES POSTAGE 3¢

The Western Pygmy Blue prefers lowland habitats such as washes, alkali flats, vacant lots, and coastlines.
Host plants are from the goosefoot family.

Winged mirrors, butterflies are soaring gifts of reflection, tiny flutterings of Love.

_____

_____

_____

_____

_____

_____

_____

_____

_____

_____

_____

_____

"When the tiny wings of the last Xerces Blue butterfly ceased to flutter, our World grew quieter by a whisper and duller by a hue." MARK JEROME WALTERS

This small periwinkle butterfly, or Flying Pansy, native to San Francisco, California, became extinct by loss of habitat.

Xerces Blue

Never more will a flying flower drift by you unnoticed.

Flower your Heart

Actual size: 1-1¼"

Never-ending rays of long summer days, violet meadows serene in my butterfly dream.

X x

PRACTICE THE LETTER 'X'

We flutter Nature's golden fields,
many gifts She yields; we gather nectar sweets,
Her blooms we leave as treats.

Lonely is the flower
And sad noontime hour
Breeze gives no pardon
Nor a longing garden
Wistful nectar and dew
Winged companion's blue
Color drops from sky
When butterfly
does not fly

Illustration after the 'Book of Hours'. From Flemish Book of Hours (c.1500)

Flower your Heart

New perspectives come to light on the wings of self-reflection ... flower your heart and bloom compassion.

Kisses by Angels

ellow Angled-Sulphur

Bloom truth, shine brightly.

ake wing ~ unfold your heart. Carry kindness ~ be transformed. Spread Grace ~ change the world.

Actual size: 2⅜–3½"

FAMILY *Pieridae*

PRACTICE BY TRACING OVER THE LETTERS

In the garden of creation, I choose elation.
As butterflies bask in warmth of days ...
with joy of the sun, I gather golden rays.

Yy

Butterflies are kisses blown by Angels.

Found near their host plant, cassia, within the southernmost states, these Yellow Angled-Sulphur butterflies glide like a bit of harnessed sunshine in flight under a clear blue sky.

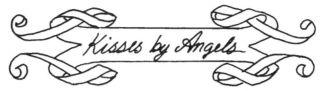

Butterfly flutters tickle the drowsy heart, awakening imaginations, curiosities, wonders, and dreams. Where will you fly once touched by a Butterfly wing?

_____

_____

_____

_____

_____

_____

_____

_____

_____

_____

_____

_____

Paint something beautiful!

Zebra Swallowtail

Actual size is 2-1/4"

With sword-like tails falling from black-and-white scales, the Zebra Swallowtail is an exotic stallion in flight, a sharp-winged dresser designed to keep predators at bay.

FAMILY Papilionidae

PRACTICE BY TRACING OVER THE LETTERS

Z Z Z z

PRACTICE THE LETTER `z`

The echo of your thoughts colors the world around you; paint something beautiful. ✿

Post Card

In many mediums does the creative soul dabble, but the most divine artistry is crafted with love.

What would the portrait of your life look like? What story would the canvas tell about you?

Dance in shades of everyhue... Let compassion color the heart of you.

Paint Something Beautiful

When infinite are the Creator's hues,
to paint black & white is to paint too few. ✿

_____
_____
_____
_____
_____
_____
_____
_____
_____
_____
_____
_____
_____

Dedication

To the Gentle One, Creator of Heaven and Earth,
who so graciously moves our spirits to Love.

For the blessing of my parents, husband, children, family, and friends,
without whom I might not have found my wings to soar.
I am ever grateful for your gentle breezes and kind affections.

With special appreciation to Maryjo Koch for bringing the beauty
of butterflies to her brush for all to be inspired by.

God unfolds our lives as He does the wings of a flying flower.
Grace rises in our hearts as butterflies from the Garden.

 **GIRL FRIDAY BOOKS**

Published by Girl Friday Books™, Seattle
www.girlfridaybooks.com

ISBN 978-1-954854-75-8
Library of Congress Control Number: 2022908016

Text by Kristen D'Angelo, www.thewhitebutterflyfund.com
Artwork by Maryjo Koch, www.kochstudios.com
Produced by Jennifer Barry, www.jenniferbarrydesign.com

For information about butterflies and butterfly gardens,
visit www.obsessionwithbutterflies.com.

To purchase butterfly gifts and garden seed packets,
visit www.crossbreezecharities.org.

To hear and stream butterfly-inspired music,
visit www.gracewing.org.

First edition